Preface

The world of technology is ever-evolving, and the Internet of Things (IoT) stands at the forefront of this transformation. As I embarked on the journey of writing "Understanding Internet of Things [IoT]," I aimed to create a resource that demystifies the complexities of IoT, making it accessible to everyone, from beginners to professionals. This book is a culmination of countless hours of research, reflection, and writing, and I hope it serves as a valuable guide for all who seek to understand this fascinating field.

I would like to express my deepest gratitude to my family, whose unwavering support has been my pillar of strength throughout this endeavour. Especially to my mother, who had always been my guiding light and source of wisdom—thank you for your endless encouragement and love. To my daughter, whose curiosity and enthusiasm for learning inspire me every day—you are my motivation and joy.

I am also deeply grateful to all my readers. Your interest in this subject is what drives me to share my knowledge and insights. I hope this book provides you with a clear and comprehensive understanding of IoT and its impact on our world. Your engagement and feedback are invaluable, and I am honoured to have you as part of this journey.

Thank you for choosing to explore the world of IoT with me.

Sincerely,
Sandeep Saini

UNDERSTANDING THE INTERNET OF THINGS (IOT)

Table of Contents

1. Introduction to IoT
2. History and Evolution of IoT
3. Key Components of IoT
4. IoT Communication Protocols
5. IoT Security and Privacy
6. Applications of IoT in Various Industries
7. Future Trends and Developments in IoT
8. Case Studies of IoT Implementations
9. Challenges and Solutions in IoT
10. IoT Development Platforms and Tools
11. IoT Standards and Protocols
12. The Future of IoT
13. IoT Case Studies
14. Ethical and Societal Implications of IoT
15. Summary and Future Directions
16. Emerging Trends and Technologies in IoT

Book Introduction

Introduction to Understanding the Internet of Things (IoT)

The Internet of Things (IoT) is a term that has rapidly gained traction in recent years, encapsulating a revolutionary concept that is reshaping industries, economies, and everyday life. At its core, IoT refers to a network of physical objects—devices, vehicles, appliances, and other items—embedded with sensors, software, and other technologies with the ability to connect and exchange data with other devices and systems over the internet.

A World Interconnected

Imagine a world where your alarm clock not only wakes you up but also signals your coffee maker to start brewing your favourite blend, where your refrigerator can order groceries when supplies are running low, and where traffic lights dynamically change to optimize flow based on real-time traffic conditions. This interconnectedness extends beyond the home into various sectors, including healthcare, agriculture, transportation, and industry, promising enhanced efficiency, reduced waste, and improved quality of life.

From Concept to Reality

The concept of interconnected devices dates back to the early days of computing and communication technologies, but it is only in recent decades that advances in miniaturization, wireless communication, and data processing have brought IoT from a futuristic vision to a tangible reality. The convergence of affordable sensors, widespread internet access, and powerful data analytics has been pivotal in this transformation.

The Scope of IoT

The scope of IoT is vast and continually expanding. According to various studies, the number of connected devices is expected to reach tens of billions in the coming years. This exponential growth is driven by the proliferation of smart devices and the increasing demand for automation and data-driven decision-making. IoT encompasses a broad spectrum of applications, from simple consumer gadgets to complex industrial systems, making it a cornerstone of the modern digital landscape.

Key Components

Understanding IoT requires a grasp of its fundamental components: sensors and actuators, connectivity, data processing, and user interfaces. Sensors collect data from the physical environment, actuators perform actions based on processed data, connectivity solutions enable communication between devices, and data processing units analyze and interpret the data to provide actionable insights. Each component plays a critical role in the functioning of IoT systems.

Communication Protocols

A variety of communication protocols facilitate the seamless exchange of data in IoT networks. These include traditional internet protocols, as well as specialized protocols designed for low-power, low-latency communication. The choice of protocol depends on the specific requirements of the application, such as range, power consumption, and data throughput.

Security and Privacy

As with any technology that handles vast amounts of data, security and privacy are paramount concerns in IoT. The interconnected nature of IoT systems creates numerous potential points of vulnerability, making them attractive targets for cyber attacks. Ensuring robust security measures and safeguarding user privacy are critical challenges that need to be addressed to build trust in IoT solutions.

Diverse Applications

IoT's potential is perhaps most vividly illustrated through its diverse applications. In smart homes, IoT enhances convenience and energy efficiency. In healthcare, it enables remote monitoring and personalized treatment. In agriculture, it optimizes resource use and boosts yields. Industrial IoT (IIoT) revolutionizes manufacturing processes, and in transportation, IoT improves safety and efficiency. Each application area brings unique benefits and challenges, demonstrating the transformative power of IoT.

Future Prospects

The future of IoT is promising but also fraught with challenges. As the number of connected devices continues to grow, issues such as interoperability, scalability, and sustainability will need to be addressed. Standardization efforts and regulatory frameworks will play crucial roles in ensuring the seamless integration and safe deployment of IoT technologies. Moreover, the ethical implications of IoT, including data ownership, consent, and the digital divide, must be carefully considered.

Chapter 1: Introduction to IoT

The term "Internet of Things" (IoT) might sound like something out of a science fiction novel, but it's a reality that is increasingly shaping our everyday lives. IoT refers to a network of physical objects—devices, vehicles, buildings, and other items—embedded with electronics, software, sensors, and network connectivity that enable these objects to collect and exchange data. In this chapter, we will explore the basics of IoT, its origins, and the fundamental technologies that make it possible.

1.1 The Birth of IoT

The concept of IoT can be traced back to the early days of the internet. However, the term itself was coined by Kevin Ashton in 1999 during his work at Procter & Gamble, where he used it to describe a system where the internet is connected to the physical world via ubiquitous sensors. Ashton's vision was to create a world where computers could gather data without human intervention, thereby creating a more efficient and interconnected world.

1.2 Defining IoT

IoT is a broad and often nebulous concept, but at its core, it is about connecting devices to the internet and to each other. This connectivity allows devices to send and receive data, enabling them to interact with the environment, perform tasks autonomously, and provide users with valuable insights. The key components of IoT include sensors, actuators, connectivity, and data processing.

- **Sensors:** These are the eyes and ears of IoT. They collect data from the physical environment, such as temperature, humidity, motion, and light. Sensors can range from simple temperature gauges to complex imaging devices.

- **Actuators:** These are the hands and feet of IoT. They perform actions based on the data collected by sensors. For example, a thermostat might adjust the heating in a home based on temperature readings, or a robotic arm might assemble a product in a factory.
- **Connectivity:** This enables devices to communicate with each other and with centralized systems. Connectivity can be achieved through various means, including Wi-Fi, Bluetooth, cellular networks, and specialized IoT protocols like Zigbee and LoRaWAN.
- **Data Processing:** This is the brain of IoT. Data collected by sensors is often processed locally on the device or in the cloud to extract meaningful insights and trigger actions. Data processing can involve simple rules-based algorithms or complex machine learning models.

1.3 The Ecosystem of IoT

The IoT ecosystem is vast and includes various stakeholders, each playing a crucial role in its development and deployment. These stakeholders include device manufacturers, software developers, network providers, and end-users. The interplay between these stakeholders determines the success and scalability of IoT solutions.

1.4 Real-World Applications

IoT is not just a theoretical concept; it is already being used in a wide range of applications. Here are a few examples:

- **Smart Homes:** IoT enables homeowners to control lighting, heating, security systems, and appliances remotely. Smart thermostats, for instance, can learn users' preferences and adjust temperatures accordingly, leading to energy savings.
- **Healthcare:** IoT devices can monitor patients' vital signs in real time, allowing for early detection of potential health issues and enabling remote care. Wearable devices like fitness trackers are also a part of this ecosystem.

- **Agriculture:** IoT can help farmers monitor soil conditions, track livestock, and optimize water usage. This leads to increased productivity and resource efficiency.
- **Industrial IoT (IIoT):** In manufacturing, IoT is used for predictive maintenance, quality control, and optimizing production processes. Connected machines can communicate with each other to streamline operations and reduce downtime.
- **Transportation:** IoT can enhance the efficiency and safety of transportation systems. Connected cars can provide real-time traffic information, while smart logistics can optimize delivery routes and reduce fuel consumption.

1.5 The Impact of IoT

The impact of IoT is profound and far-reaching. It has the potential to revolutionize industries, improve quality of life, and create new business opportunities. However, it also presents challenges, such as security vulnerabilities and privacy concerns. Addressing these challenges is crucial to realizing the full potential of IoT.

1.6 The Future of IoT

The future of IoT looks promising, with continued advancements in technology and increasing adoption across various sectors. Emerging trends such as 5G connectivity, edge computing, and artificial intelligence (AI) are expected to further enhance the capabilities of IoT. As the number of connected devices grows, the potential for IoT to transform our world becomes even greater.

Chapter 2: History and Evolution of IoT

2.1 Early Beginnings

The roots of IoT can be traced back to the early days of computing and communication technologies. The invention of the internet in the late 20th century laid the groundwork for the interconnected world we live in today. The first recognizable IoT device was a Coca-Cola vending machine at Carnegie Mellon University in the early 1980s, which was connected to the internet and could report its inventory and whether newly loaded drinks were cold.

2.2 The 1990s: Birth of IoT Concept

The 1990s saw significant advancements in technology, including the development of RFID (Radio-Frequency Identification) technology. Kevin Ashton, a British technology pioneer, coined the term "Internet of Things" in 1999 while working on RFID at Procter & Gamble. Ashton's vision was to connect the physical world to the internet through a network of sensors, allowing computers to gather data without human intervention.

2.3 The 2000s: Growth and Development

The early 2000s witnessed the growth of IoT with the advent of ubiquitous computing. Advances in wireless technology, miniaturization of sensors, and improvements in data processing capabilities contributed to the development of IoT devices. The introduction of IPv6 in 1999 also played a crucial role by providing a virtually unlimited number of IP addresses, essential for the proliferation of connected devices.

2.4 The 2010s: IoT Goes Mainstream

The 2010s marked the mainstream adoption of IoT. The launch of smart home devices like the Nest thermostat and the Amazon Echo brought IoT into the consumer spotlight. Industrial IoT (IIoT) also gained traction, with companies leveraging IoT for predictive maintenance, asset tracking, and process optimization. The rise of cloud computing provided the necessary infrastructure for handling the massive amounts of data generated by IoT devices.

2.5 Key Milestones in IoT Evolution

- **2008-2009:** The number of connected devices surpassed the global population.
- **2011:** The International Telecommunication Union (ITU) officially recognized IoT as a key emerging technology.
- **2013:** Google acquired Nest Labs for $3.2 billion, signalling the importance of IoT in the tech industry.
- **2014:** The Industrial Internet Consortium (IIC) was founded to accelerate the growth of IIoT.
- **2016:** The Mirai botnet attack highlighted the security vulnerabilities of IoT devices, leading to increased focus on IoT security.

2.6 Current State of IoT

Today, IoT is an integral part of our lives, with applications ranging from smart homes and healthcare to agriculture and transportation. The number of connected devices is projected to reach 75 billion by 2025, driven by advancements in technology and increasing demand for automation and data-driven decision-making.

2.7 Future Trends in IoT

The future of IoT looks promising, with several emerging trends poised to shape its evolution:

- **5G Connectivity:** The rollout of 5G networks will provide faster and more reliable connectivity, enabling real-time communication between IoT devices.
- **Edge Computing:** Moving data processing closer to the source (at the edge) will reduce latency and bandwidth usage, enhancing the performance of IoT applications.
- **Artificial Intelligence (AI):** AI will play a crucial role in analyzing the vast amounts of data generated by IoT devices, providing actionable insights and enabling autonomous decision-making.
- **Blockchain:** Blockchain technology can enhance the security and transparency of IoT systems by providing a decentralized and tamper-proof ledger for recording transactions.
- **Interoperability:** Efforts to standardize IoT protocols and ensure interoperability between devices will be crucial for the seamless integration of IoT solutions.

Chapter 3: Key Components of IoT

3.1 Sensors and Actuators

At the heart of IoT systems are sensors and actuators. Sensors gather data from the environment, such as temperature, humidity, light, motion, and pressure. These small devices convert physical parameters into electrical signals. Actuators, on the other hand, receive signals from a controller and perform actions like turning on a light, opening a valve, or moving a robotic arm. Together, sensors and actuators enable IoT devices to interact with the physical world.

- **Types of Sensors:**
    - **Temperature Sensors:** Measure heat energy and convert it to readable data.
    - **Motion Sensors:** Detect physical movement in an area.
    - **Proximity Sensors:** Determine the presence of objects without physical contact.
    - **Light Sensors:** Measure light intensity, commonly used in smart phones and smart lighting systems.
    - **Humidity Sensors:** Measure the amount of water vapour in the air.
    - **Pressure Sensors:** Measure the force applied to a surface.

3.2 Connectivity

Connectivity is a crucial component of IoT, allowing devices to communicate with each other and with centralized systems. There are various connectivity options, each suited for different applications and requirements.

- **Wi-Fi:** Commonly used for home and office IoT devices due to its high data rate and widespread availability.

- **Bluetooth:** Ideal for short-range communication, often used in wearable devices and smart home products.
- **Zigbee:** A low-power, low-data-rate wireless network standard, suitable for applications like smart lighting and home automation.
- **LoRaWAN:** A long-range, low-power wide-area network protocol, used for applications like agriculture and environmental monitoring.
- **Cellular Networks:** Provide wide coverage and high reliability, used for mobile IoT devices and applications in remote areas.
- **NB-IoT:** A narrowband IoT technology designed for low-power devices with long battery life, used in smart metering and asset tracking.

3.3 Data Processing and Analytics

Data collected by sensors is only valuable if it can be processed and analyzed to extract meaningful insights. This involves several steps, including data collection, storage, processing, and analysis.

- **Edge Computing:** Data is processed close to the source (at the edge) rather than in a centralized cloud, reducing latency and bandwidth usage.
- **Cloud Computing:** Provides scalable storage and processing power for large volumes of IoT data. Cloud platforms offer various services, from data storage to advanced analytics.
- **Artificial Intelligence (AI) and Machine Learning (ML):** These technologies analyze IoT data to identify patterns, make predictions, and enable autonomous decision-making. For example, AI can predict equipment failures based on sensor data, reducing downtime in industrial settings.

3.4 User Interfaces

User interfaces (UIs) are essential for interacting with IoT systems. They provide users with a means to monitor and control IoT devices.

- **Mobile Apps:** Commonly used for smart home devices, allowing users to control lighting, security systems, and appliances from their smart phones.
- **Web Dashboards:** Provide a centralized view of IoT data, often used in industrial and commercial applications to monitor systems and generate reports.
- **Voice Assistants:** Devices like Amazon Echo and Google Home use voice commands to control IoT devices, enhancing user convenience.
- **Wearables:** Smart watches and fitness trackers provide users with real-time data on their health and activities.

3.5 Power Management

Power management is a critical aspect of IoT, particularly for battery-powered devices. Efficient power management ensures long battery life and reliable operation.

- **Energy Harvesting:** Techniques like solar power, vibration energy, and thermal energy are used to generate power for IoT devices, reducing dependence on batteries.
- **Low-Power Designs:** IoT devices are designed to consume minimal power, using low-power communication protocols and energy-efficient components.
- **Battery Technologies:** Advances in battery technology, such as longer-lasting lithium-ion batteries, are essential for sustaining IoT devices over extended periods.

Chapter 4: IoT Communication Protocols

4.1 Overview of IoT Communication Protocols

Communication protocols are the backbone of IoT, enabling devices to exchange data seamlessly. These protocols define the rules and conventions for data transmission, ensuring reliable and efficient communication. The choice of protocol depends on various factors, including range, data rate, power consumption, and application requirements.

4.2 Common IoT Communication Protocols

- **MQTT (Message Queuing Telemetry Transport):** A lightweight protocol designed for low-bandwidth, high-latency networks. MQTT is ideal for applications like remote monitoring and control, where bandwidth efficiency is crucial.
- **CoAP (Constrained Application Protocol):** Designed for simple devices and low-power networks, CoAP is used in applications like smart lighting and environmental monitoring. It uses a request/response model similar to HTTP but with reduced overhead.
- **HTTP/HTTPS:** While not specifically designed for IoT, HTTP/HTTPS is widely used due to its ubiquity and compatibility with web technologies. It is suitable for applications where security and integration with web services are important.
- **Zigbee:** A low-power, low-data-rate wireless protocol designed for personal area networks. Zigbee is commonly used in home automation, smart lighting, and building control systems.
- **LoRaWAN (Long Range Wide Area Network):** A protocol for long-range, low-power communication. LoRaWAN is ideal for applications like smart agriculture, asset tracking, and environmental monitoring.
- **Bluetooth Low Energy (BLE):** A wireless protocol designed for short-range communication with low power consumption. BLE is

used in wearables, health monitoring devices, and smart home products.
- **NB-IoT (Narrowband IoT):** A cellular technology designed for IoT devices with low data rates and long battery life. NB-IoT is used in smart metering, asset tracking, and environmental monitoring.
- **6LoWPAN (IPv6 over Low-Power Wireless Personal Area Networks):** Allows IPv6 packets to be sent over low-power wireless networks. It is used in applications like smart grids, home automation, and industrial monitoring.

4.3 Factors Influencing Protocol Choice

- **Range:** The distance over which data needs to be transmitted. Long-range protocols like LoRaWAN and NB-IoT are suitable for wide-area applications, while short-range protocols like Bluetooth and Zigbee are used for personal area networks.
- **Data Rate:** The amount of data that needs to be transmitted. High-data-rate protocols like Wi-Fi are used for applications requiring large amounts of data, while low-data-rate protocols like MQTT are used for simple sensor data.
- **Power Consumption:** The energy efficiency of the protocol. Low-power protocols like BLE and Zigbee are suitable for battery-powered devices, while higher-power protocols like Wi-Fi are used where power availability is not a concern.
- **Latency:** The time it takes for data to be transmitted and received. Low-latency protocols are essential for real-time applications like industrial automation and telemedicine.
- **Security:** The level of security required for data transmission. Protocols like HTTPS provide strong security features, while others may require additional security measures.

4.4 Emerging Protocols and Trends

- **5G:** The fifth generation of cellular networks promises high data rates, low latency, and massive connectivity, making it suitable for a wide range of IoT applications, including smart cities and autonomous vehicles.
- **LPWAN (Low Power Wide Area Network):** Technologies like Sigfox and LoRaWAN are gaining traction for their ability to provide long-range, low-power communication, ideal for remote and rural IoT deployments.
- **Thread:** A low-power, secure, and scalable protocol designed for smart home devices. Thread supports IPv6 and is interoperable with existing networks.
- **OPC UA (Open Platform Communications Unified Architecture):** A protocol for industrial IoT, providing secure and reliable communication for industrial automation and control systems.

Chapter 5: IoT Security and Privacy

5.1 Importance of Security and Privacy in IoT

The proliferation of IoT devices has brought significant benefits but also introduced new security and privacy challenges. IoT systems often handle sensitive data, and their interconnected nature makes them attractive targets for cyber attacks. Ensuring robust security and safeguarding user privacy are critical for building trust in IoT solutions.

5.2 Common IoT Security Threats

- **Device Hijacking:** Attackers gain control of IoT devices to perform unauthorized actions, such as turning off security cameras or accessing personal data.
- **Data Breaches:** Unauthorized access to data collected by IoT devices, leading to the exposure of sensitive information like health records, financial data, and personal details.
- **Distributed Denial of Service (DDoS) Attacks:** Compromised IoT devices are used to launch large-scale attacks on network infrastructure, disrupting services and causing significant damage.
- **Firmware Vulnerabilities:** Weaknesses in the software running on IoT devices can be exploited by attackers to gain unauthorized access or disrupt device functionality.
- **Privacy Invasion:** Unauthorized access to personal data collected by IoT devices, leading to privacy violations and potential misuse of information.

5.3 Security Best Practices for IoT

- **Device Authentication:** Ensuring that only authorized devices can connect to the IoT network. This can be achieved through

techniques like digital certificates, two-factor authentication, and secure boot processes.
- **Data Encryption:** Encrypting data both in transit and at rest to protect it from unauthorized access. Using strong encryption protocols like AES (Advanced Encryption Standard) ensures data confidentiality.
- **Regular Updates and Patch Management:** Keeping IoT device firmware and software up to date with the latest security patches to address vulnerabilities and protect against emerging threats.
- **Network Segmentation:** Isolating IoT devices from other network components to limit the potential impact of a compromised device. This can be achieved through virtual LANs (VLANs) and firewalls.
- **Security by Design:** Incorporating security measures into the design and development of IoT devices and systems, rather than as an afterthought. This includes secure coding practices, regular security testing, and adherence to industry standards.

5.4 Privacy Considerations in IoT

- **Data Minimization:** Collecting only the data necessary for the intended purpose and avoiding excessive data collection. This reduces the risk of privacy breaches and minimizes the amount of sensitive information that needs protection.
- **User Consent:** Obtaining explicit consent from users before collecting, processing, or sharing their data. Transparent privacy policies and user-friendly consent mechanisms help build trust with users.
- **Anonymization and Pseudonymization:** Techniques like anonymization and pseudonymization help protect user privacy by removing or masking personally identifiable information (PII) from collected data.
- **Data Ownership and Control:** Ensuring that users retain ownership and control over their data. Providing users with options

to access, modify, or delete their data enhances privacy and compliance with data protection regulations.

5.5 Regulatory Compliance

- **GDPR (General Data Protection Regulation):** The European Union's GDPR sets strict requirements for data protection and privacy, including the rights of individuals to access and control their data. IoT solutions deployed in the EU must comply with GDPR regulations.
- **CCPA (California Consumer Privacy Act):** The CCPA provides California residents with rights related to their personal data, including the right to know what data is collected, the right to delete data, and the right to opt-out of data selling.
- **HIPAA (Health Insurance Portability and Accountability Act):** IoT devices used in healthcare must comply with HIPAA regulations, which protect the privacy and security of health information.

Chapter 6: Applications of IoT in Various Industries

6.1 Smart Homes

The concept of a smart home involves the use of IoT devices to automate and enhance the living environment. Smart home applications include:

- **Home Automation:** IoT devices like smart thermostats, lighting systems, and security cameras provide convenience and energy efficiency by automating routine tasks and allowing remote control via mobile apps.
- **Energy Management:** Smart meters and energy monitoring systems help homeowners track energy consumption and identify ways to reduce usage, leading to cost savings and environmental benefits.
- **Home Security:** IoT-enabled security systems, including smart locks, video doorbells, and motion sensors, enhance home security by providing real-time alerts and remote monitoring capabilities.

6.2 Healthcare

IoT has revolutionized healthcare by enabling remote monitoring, personalized treatment, and improved patient outcomes. Key applications include:

- **Wearable Health Devices:** Devices like fitness trackers, smart watches, and medical wearables monitor vital signs, physical activity, and other health metrics, providing valuable data for healthcare providers and individuals.
- **Remote Patient Monitoring:** IoT devices enable continuous monitoring of patients' health conditions, allowing for early detection of issues and reducing the need for frequent hospital visits.

- **Smart Medical Equipment:** IoT-enabled medical devices, such as connected inhalers and insulin pumps, provide real-time data to healthcare providers, improving treatment and management of chronic conditions.

6.3 Industrial IoT (IIoT)

The industrial sector leverages IoT to optimize operations, improve efficiency, and enhance safety. Key IIoT applications include:

- **Predictive Maintenance:** IoT sensors monitor equipment health and performance, enabling predictive maintenance to prevent breakdowns and reduce downtime.
- **Asset Tracking:** IoT devices track the location and condition of assets, improving inventory management and logistics.
- **Process Automation:** IoT systems automate industrial processes, enhancing productivity and reducing operational costs.

6.4 Smart Cities

IoT plays a crucial role in developing smart cities by improving urban living through efficient resource management and enhanced services. Key applications include:

- **Smart Traffic Management:** IoT-enabled traffic sensors and intelligent transportation systems optimize traffic flow, reduce congestion, and improve safety.
- **Waste Management:** Smart bins equipped with sensors monitor waste levels and optimize collection routes, reducing operational costs and environmental impact.
- **Environmental Monitoring:** IoT devices monitor air quality, water quality, and noise levels, providing valuable data for urban planning and environmental protection.

6.5 Agriculture

IoT is transforming agriculture by enabling precision farming and improving crop yield and resource management. Key applications include:

- **Precision Farming:** IoT sensors monitor soil conditions, weather, and crop health, providing data for optimizing irrigation, fertilization, and pest control.
- **Livestock Monitoring:** IoT devices track the health and behaviour of livestock, enabling early detection of diseases and improving animal welfare.
- **Smart Irrigation:** IoT-enabled irrigation systems adjust water usage based on real-time data, reducing water waste and improving crop yield.

6.6 Transportation and Logistics

IoT enhances transportation and logistics by improving efficiency, safety, and customer service. Key applications include:

- **Fleet Management:** IoT devices track the location, condition, and performance of vehicles, optimizing routes, reducing fuel consumption, and improving maintenance.
- **Supply Chain Monitoring:** IoT sensors monitor the condition of goods during transit, ensuring quality and reducing losses.
- **Autonomous Vehicles:** IoT technologies enable the development of self-driving cars, enhancing safety and transforming the transportation industry.

6.7 Retail

The retail industry leverages IoT to enhance the shopping experience, streamline operations, and improve inventory management. Key applications include:

- **Smart Shelves:** IoT-enabled shelves monitor product availability and automatically update inventory, reducing stock outs and improving customer satisfaction.
- **Personalized Shopping:** IoT devices track customer preferences and behaviour, enabling personalized recommendations and targeted marketing.
- **Supply Chain Optimization:** IoT sensors monitor the condition and location of products throughout the supply chain, ensuring timely delivery and reducing losses.

Chapter 7: Future Trends and Developments in IoT

7.1 5G Connectivity and IoT

The rollout of 5G networks is set to revolutionize IoT by providing faster, more reliable connectivity with lower latency. Key benefits of 5G for IoT include:

- **Enhanced Speed:** 5G offers significantly higher data rates, enabling real-time communication and faster data transfer between IoT devices.
- **Low Latency:** Reduced latency ensures quick response times, critical for applications like autonomous vehicles and industrial automation.
- **Massive Connectivity:** 5G can support a vast number of connected devices, making it ideal for large-scale IoT deployments in smart cities and industrial settings.

7.2 Edge Computing and IoT

Edge computing involves processing data closer to the source (at the edge) rather than in a centralized cloud. Key benefits of edge computing for IoT include:

- **Reduced Latency:** Processing data at the edge reduces latency, enhancing the performance of real-time IoT applications.
- **Bandwidth Efficiency:** By processing data locally, edge computing reduces the amount of data that needs to be transmitted to the cloud, optimizing bandwidth usage.
- **Enhanced Security:** Edge computing can improve security by keeping sensitive data closer to the source and reducing exposure to potential cyber threats.

7.3 Artificial Intelligence and Machine Learning

AI and ML are set to play a crucial role in the future of IoT, enabling smarter, more autonomous systems. Key applications of AI and ML in IoT include:

- **Predictive Analytics:** AI algorithms analyze IoT data to predict future events, such as equipment failures or demand fluctuations, enabling proactive decision-making.
- **Autonomous Systems:** Machine learning enables IoT devices to learn from data and make autonomous decisions, enhancing efficiency and reducing human intervention.
- **Personalization:** AI-powered IoT devices can provide personalized experiences by learning from user behaviour and preferences.

7.4 Block chain and IoT

Block chain technology can enhance the security and transparency of IoT systems by providing a decentralized and tamper-proof ledger for recording transactions. Key benefits of Blockchain for IoT include:

- **Enhanced Security:** Blockchain's decentralized nature makes it difficult for attackers to alter data, providing a secure foundation for IoT applications.
- **Transparency:** Blockchain enables transparent and auditable records of IoT transactions, enhancing trust and accountability.
- **Smart Contracts:** Blockchain supports smart contracts, enabling automated and secure interactions between IoT devices and systems.

7.5 IoT and Sustainability

IoT has the potential to drive sustainability efforts by optimizing resource usage and reducing environmental impact. Key applications of IoT for sustainability include:

- **Energy Management:** IoT devices monitor and optimize energy usage in buildings, reducing consumption and promoting energy efficiency.
- **Smart Agriculture:** IoT-enabled precision farming techniques reduce water and fertilizer usage, enhancing sustainability in agriculture.
- **Environmental Monitoring:** IoT sensors monitor environmental conditions, providing data for informed decision-making and environmental protection.

Chapter 8: Case Studies of IoT Implementations

The real-world applications of IoT are vast and varied, providing valuable insights into how this technology is transforming different sectors. This chapter delves into specific case studies that highlight successful IoT implementations across various industries. These examples illustrate the practical benefits, challenges, and innovative solutions associated with IoT deployments.

8.1 Smart Home Case Study: Nest Thermostat

Overview: The Nest Thermostat developed by Nest Labs (acquired by Google), is a prime example of IoT in the smart home sector. Launched in 2011, the Nest Thermostat revolutionized home energy management by learning users' habits and automatically adjusting temperatures for optimal energy efficiency.

Key Features:

- **Learning Capability:** The Nest Thermostat learns user preferences over time and adjusts heating and cooling schedules accordingly.
- **Remote Control:** Users can control the thermostat remotely via a mobile app, allowing adjustments from anywhere.
- **Energy Reports:** Monthly energy reports provide insights into energy usage patterns, helping users make informed decisions to save energy.

Benefits:

- **Energy Savings:** The Nest Thermostat has been shown to save users up to 15% on cooling and 12% on heating bills.

- **Convenience:** Automation and remote control features provide significant convenience for users.
- **Environmental Impact:** Reduced energy consumption contributes to lower greenhouse gas emissions.

Challenges:

- **Privacy Concerns:** As with many IoT devices, there are concerns about data privacy and the security of user information.
- **Compatibility:** Ensuring compatibility with various HVAC systems can be a challenge.

8.2 Healthcare Case Study: Remote Patient Monitoring

Overview: Remote patient monitoring (RPM) uses IoT devices to collect and transmit health data from patients to healthcare providers. One notable example is the use of IoT in managing chronic conditions like diabetes and heart disease.

Key Features:

- **Continuous Monitoring:** IoT devices such as glucose monitors and heart rate monitors continuously collect health data.
- **Data Transmission:** Collected data is transmitted in real-time to healthcare providers for analysis and intervention.
- **Alerts and Notifications:** Automatic alerts are sent to patients and providers if readings fall outside of predefined ranges.

Benefits:

- **Improved Patient Outcomes:** Continuous monitoring allows for early detection of potential issues, leading to timely interventions.

- **Reduced Hospital Visits:** RPM reduces the need for frequent hospital visits, improving patient convenience and reducing healthcare costs.
- **Personalized Care:** Data-driven insights enable personalized treatment plans tailored to individual patient needs.

Challenges:

- **Data Security:** Ensuring the security of sensitive health data is a significant concern.
- **Device Reliability:** The accuracy and reliability of IoT devices are critical for effective monitoring.

8.3 Industrial IoT Case Study: Predictive Maintenance in Manufacturing

Overview: Predictive maintenance (PdM) uses IoT sensors and data analytics to predict equipment failures before they occur, enabling proactive maintenance and reducing downtime. General Electric (GE) has successfully implemented PdM in its manufacturing operations.

Key Features:

- **Sensor Integration:** IoT sensors are installed on machinery to monitor parameters such as vibration, temperature, and pressure.
- **Data Analytics:** Collected data is analyzed using advanced analytics and machine learning algorithms to predict potential failures.
- **Maintenance Scheduling:** Maintenance is scheduled based on predictive insights, preventing unexpected breakdowns.

Benefits:

- **Reduced Downtime:** PdM minimizes unplanned downtime, improving overall operational efficiency.
- **Cost Savings:** Preventive maintenance reduces repair costs and extends the lifespan of equipment.
- **Increased Safety:** Identifying and addressing issues before they lead to failures enhances workplace safety.

Challenges:

- **Implementation Cost:** Initial costs for sensors, analytics software, and integration can be high.
- **Data Management:** Managing and analyzing large volumes of data requires robust infrastructure and expertise.

8.4 Smart City Case Study: Barcelona's Smart City Initiatives

Overview: Barcelona has emerged as a global leader in smart city initiatives, leveraging IoT to enhance urban living and sustainability. The city's smart infrastructure includes smart lighting, waste management, and traffic systems.

Key Features:

- **Smart Lighting:** IoT-enabled streetlights adjust brightness based on real-time conditions, reducing energy consumption.
- **Waste Management:** Smart bins equipped with sensors monitor waste levels and optimize collection routes.
- **Traffic Management:** IoT sensors and intelligent transportation systems manage traffic flow and reduce congestion.

Benefits:

- **Energy Efficiency:** Smart lighting and waste management systems reduce energy usage and operational costs.
- **Improved Services:** Enhanced traffic management leads to reduced congestion and improved public transportation.
- **Sustainability:** Smart city initiatives contribute to environmental sustainability by optimizing resource usage.

Challenges:

- **Data Privacy:** Ensuring the privacy and security of data collected from various sources is a critical concern.
- **Integration:** Integrating multiple IoT systems and ensuring interoperability can be complex.

8.5 Agricultural IoT Case Study: Precision Farming with John Deere

Overview: John Deere, a leading manufacturer of agricultural machinery, has integrated IoT into its equipment to enable precision farming. This approach optimizes farming practices and improves crop yields.

Key Features:

- **IoT Sensors:** Sensors monitor soil conditions, weather, and crop health in real-time.
- **Data Analytics:** Collected data is analyzed to provide actionable insights for irrigation, fertilization, and pest control.
- **Automated Equipment:** IoT-enabled machinery performs tasks with precision based on data-driven recommendations.

Benefits:

- **Increased Crop Yield:** Precision farming techniques result in higher crop yields and better quality produce.
- **Resource Efficiency:** Optimizing water, fertilizer, and pesticide usage reduces waste and environmental impact.
- **Cost Savings:** Efficient resource usage leads to cost savings for farmers.

Challenges:

- **Technology Adoption:** Farmers need to adopt new technologies and invest in IoT-enabled equipment.
- **Data Management:** Managing and analyzing large volumes of data requires technical expertise and infrastructure.

8.6 Transportation Case Study: UPS and IoT in Logistics

Overview: UPS has leveraged IoT to optimize its logistics operations, improving efficiency and customer service. The company's ORION (On-Road Integrated Optimization and Navigation) system uses IoT data for route optimization.

Key Features:

- **Real-Time Tracking:** IoT sensors track the location and condition of packages throughout the delivery process.
- **Route Optimization:** The ORION system uses real-time data to optimize delivery routes, reducing travel time and fuel consumption.
- **Predictive Maintenance:** IoT-enabled vehicles are monitored for maintenance needs, preventing breakdowns and delays.

Benefits:

- **Operational Efficiency:** Route optimization and predictive maintenance improve operational efficiency and reduce costs.
- **Customer Satisfaction:** Real-time tracking and timely deliveries enhance customer satisfaction.
- **Environmental Impact:** Optimized routes reduce fuel consumption and greenhouse gas emissions.

Challenges:

- **Data Security:** Ensuring the security of sensitive shipment data is essential.
- **System Integration:** Integrating IoT systems with existing logistics infrastructure can be complex.

8.7 Retail Case Study: Amazon Go and IoT in Retail

Overview: Amazon Go stores leverage IoT and AI to create a checkout-free shopping experience. Customers can enter the store, pick up items, and leave without waiting in line or interacting with a cashier.

Key Features:

- **IoT Sensors and Cameras:** Sensors and cameras track items picked up by customers and automatically add them to the virtual cart.
- **Mobile App Integration:** Customers use the Amazon Go app to enter the store and receive a digital receipt after leaving.
- **AI and Machine Learning:** AI algorithms analyze data from sensors and cameras to ensure accurate item tracking and billing.

Benefits:

- **Convenience:** The checkout-free experience provides unparalleled convenience for customers.
- **Reduced Wait Times:** Eliminating checkout lines reduces wait times and enhances the shopping experience.
- **Operational Efficiency:** Automation reduces the need for manual labour, lowering operational costs.

Challenges:

- **Privacy Concerns:** The extensive use of cameras and sensors raises privacy concerns among customers.
- **Technical Complexity:** Implementing and maintaining the IoT and AI infrastructure is technically challenging.

These case studies illustrate the diverse applications and benefits of IoT across various industries. From smart homes and healthcare to manufacturing and retail, IoT is driving innovation and delivering tangible value. By learning from these successful implementations, organizations can better understand how to leverage IoT for their own needs, overcoming challenges and reaping the rewards of this transformative technology.

Chapter 9: Challenges and Solutions in IoT

While the Internet of Things (IoT) offers numerous benefits and transformative potential, its implementation and operation come with several challenges. Addressing these challenges is crucial for the successful deployment and sustainable growth of IoT solutions. This chapter explores the key challenges in IoT and presents potential solutions to overcome them.

9.1 Interoperability Issues

Overview: Interoperability refers to the ability of different IoT devices and systems to work together seamlessly. Given the diversity of IoT devices, platforms, and protocols, achieving interoperability is a significant challenge.

Challenges:

- **Diverse Protocols:** IoT devices use various communication protocols, making it difficult to ensure seamless interaction.
- **Vendor Lock-In:** Different vendors may use proprietary standards, leading to compatibility issues.
- **Fragmented Ecosystem:** The IoT ecosystem is fragmented with numerous standards and lack of a unified framework.

Solutions:

- **Standardization Efforts:** Industry-wide standardization efforts, such as those by the Internet Engineering Task Force (IETF) and the Institute of Electrical and Electronics Engineers (IEEE), can promote interoperability.
- **Middleware Solutions:** Middleware can act as an intermediary layer, facilitating communication between diverse IoT devices and systems.

- **Open Standards:** Encouraging the use of open standards and protocols can reduce compatibility issues and vendor lock-in.

9.2 Scalability Challenges

Overview: As the number of connected devices grows exponentially, scalability becomes a critical concern. Ensuring that IoT systems can handle increasing data volumes and device connections without performance degradation is essential.

Challenges:

- **Data Overload:** Managing and processing large volumes of data from numerous IoT devices can overwhelm existing infrastructure.
- **Network Congestion:** Increased device connectivity can lead to network congestion and reduced performance.
- **Resource Constraints:** IoT devices often have limited computational and storage resources, complicating scalability.

Solutions:

- **Edge Computing:** Implementing edge computing can reduce the burden on central servers by processing data locally on the device or nearby.
- **Cloud Integration:** Leveraging cloud services can provide scalable storage and computing power to handle large data volumes.
- **Efficient Protocols:** Using efficient communication protocols and data compression techniques can mitigate network congestion and resource constraints.

9.3 Security and Privacy Concerns

Overview: Security and privacy are among the most critical challenges in IoT. The widespread deployment of IoT devices increases the attack surface, making systems vulnerable to cyber attacks. Protecting sensitive data and ensuring user privacy are paramount.

Challenges:

- **Data Breaches:** IoT devices can be targets for cyber attacks, leading to data breaches and unauthorized access.
- **Device Vulnerabilities:** Many IoT devices have weak security mechanisms, making them susceptible to exploitation.
- **Privacy Risks:** Collecting and transmitting personal data can raise significant privacy concerns.

Solutions:

- **Strong Encryption:** Implementing robust encryption techniques can protect data during transmission and storage.
- **Regular Updates:** Ensuring that IoT devices receive regular firmware and security updates can mitigate vulnerabilities.
- **Privacy by Design:** Incorporating privacy considerations into the design of IoT systems can help protect user data and ensure compliance with regulations.

9.4 Data Management and Analytics

Overview: Effective data management and analytics are crucial for deriving actionable insights from IoT data. Handling vast amounts of data generated by IoT devices poses significant challenges.

Challenges:

- **Data Quality:** Ensuring the accuracy, consistency, and completeness of IoT data is essential for reliable analytics.

- **Storage and Processing:** Storing and processing large volumes of data in real-time can strain existing infrastructure.
- **Complex Analytics:** Advanced analytics and machine learning models require substantial computational resources and expertise.

Solutions:

- **Data Governance:** Implementing robust data governance frameworks can ensure data quality and consistency.
- **Distributed Storage:** Using distributed storage solutions, such as Hadoop and NoSQL databases, can efficiently manage large data volumes.
- **Advanced Analytics Tools:** Leveraging advanced analytics tools and platforms can facilitate real-time data processing and complex analytics.

9.5 Regulatory and Compliance Issues

Overview: Compliance with regulatory requirements is crucial for the deployment of IoT systems. Different regions have varying regulations related to data privacy, security, and usage, complicating compliance efforts.

Challenges:

- **Varying Regulations:** Navigating different regulatory landscapes across regions can be challenging for global IoT deployments.
- **Compliance Costs:** Meeting regulatory requirements can incur significant costs and resource allocation.
- **Evolving Standards:** Keeping up with evolving regulatory standards and ensuring continuous compliance can be difficult.

Solutions:

- **Regulatory Frameworks:** Developing comprehensive regulatory frameworks that address IoT-specific issues can provide clarity and guidance.
- **Compliance Programs:** Implementing robust compliance programs and conducting regular audits can ensure adherence to regulations.
- **Collaborative Efforts:** Industry collaboration and engagement with regulatory bodies can help shape favourable regulations and standards.

Chapter 10: IoT Development Platforms and Tools

The successful implementation of IoT solutions relies heavily on robust development platforms and tools. These platforms and tools provide the necessary infrastructure, software, and services to design, develop, deploy, and manage IoT applications. This chapter explores the key IoT development platforms, popular tools for IoT prototyping, cloud platforms, and edge computing solutions that are pivotal for creating and managing IoT ecosystems.

10.1 Overview of IoT Development Platforms

Definition: IoT development platforms are integrated environments that provide a suite of tools and services for building, deploying, and managing IoT applications. They typically include device management, data management, analytics, and connectivity services.

Importance:

- **Simplified Development:** These platforms simplify the development process by providing pre-built modules and libraries.
- **Scalability:** They offer scalable solutions that can grow with the increasing number of devices and data.
- **Interoperability:** Many platforms support a wide range of protocols and standards, ensuring interoperability between different devices and systems.

10.2 Popular IoT Development Platforms

1. AWS IoT Core

Overview: AWS IoT Core is a managed cloud service that enables connected devices to interact with cloud applications and other devices. It offers secure device connectivity, data processing, and integration with other AWS services.

Key Features:

- **Secure Communication:** AWS IoT Core provides secure communication with encryption and mutual authentication.
- **Device Management:** It includes tools for managing devices throughout their lifecycle.
- **Data Processing:** Integration with AWS services like Lambda, S3, and DynamoDB allows for efficient data processing and storage.

Use Cases:

- **Smart Home Applications:** AWS IoT Core is used for developing smart home devices that interact with cloud services.
- **Industrial IoT:** It supports industrial applications like predictive maintenance and asset tracking.

2. Microsoft Azure IoT Hub

Overview: Azure IoT Hub is a managed service that acts as a central message hub for bi-directional communication between IoT applications and devices. It supports secure and reliable device connectivity and management.

Key Features:

- **Bi-Directional Communication:** Supports both device-to-cloud and cloud-to-device communication.
- **Security:** Provides enhanced security features like device authentication and secure data transmission.

- **Integration:** Seamless integration with other Azure services like Azure Machine Learning and Azure Stream Analytics.

Use Cases:

- **Healthcare:** Azure IoT Hub is used for remote patient monitoring and healthcare device management.
- **Smart Cities:** It supports smart city initiatives like smart lighting and traffic management.

3. Google Cloud IoT

Overview: Google Cloud IoT is a set of fully managed services that allow you to connect, manage, and ingest data from globally dispersed devices. It includes Cloud IoT Core, Cloud IoT Edge, and other related services.

Key Features:

- **Device Connectivity:** Secure and scalable connectivity for IoT devices.
- **Data Analytics:** Integration with Google Cloud's big data and machine learning services for advanced analytics.
- **Edge Computing:** Cloud IoT Edge extends Google Cloud's data processing capabilities to the edge.

Use Cases:

- **Retail:** Google Cloud IoT is used for inventory management and customer analytics in retail.
- **Transportation:** It supports fleet management and logistics optimization.

10.3 Tools for IoT Prototyping

1. Arduino

Overview: Arduino is an open-source electronics platform based on easy-to-use hardware and software. It is widely used for prototyping and developing simple IoT projects.

Key Features:

- **Versatility:** Supports a wide range of sensors and actuators.
- **Ease of Use:** User-friendly development environment and extensive community support.
- **Cost-Effective:** Affordable development boards and components.

Use Cases:

- **DIY Projects:** Popular for hobbyist and educational IoT projects.
- **Prototyping:** Used for quick prototyping and proof-of-concept development.

2. Raspberry Pi

Overview: Raspberry Pi is a small, affordable computer that can be used for various programming and electronics projects, including IoT applications.

Key Features:

- **Computational Power:** More powerful than microcontroller-based platforms like Arduino.
- **Connectivity:** Built-in networking capabilities for IoT projects.
- **Flexibility:** Can run various operating systems and supports multiple programming languages.

Use Cases:

- **Home Automation:** Used for developing smart home applications and automation systems.
- **Edge Computing:** Suitable for edge computing applications due to its processing capabilities.

10.4 Cloud Platforms for IoT

1. IBM Watson IoT

Overview: IBM Watson IoT is a cloud-based platform that provides a range of services for connecting, managing, and analyzing IoT devices and data.

Key Features:

- **Advanced Analytics:** Integration with IBM Watson's AI and machine learning services.
- **Device Management:** Tools for managing and monitoring IoT devices.
- **Scalability:** Scalable infrastructure for handling large volumes of IoT data.

Use Cases:

- **Industrial IoT:** Used for predictive maintenance and operational optimization in manufacturing.
- **Smart Buildings:** Supports building automation and energy management solutions.

2. Oracle IoT Cloud

Overview: Oracle IoT Cloud is a comprehensive IoT platform that enables organizations to connect, analyze, and integrate IoT data with enterprise applications.

Key Features:

- **Integration:** Seamless integration with Oracle's enterprise software suite.
- **Data Visualization:** Tools for visualizing and analyzing IoT data.
- **Security:** Robust security features for data protection and device management.

Use Cases:

- **Supply Chain Management:** Used for tracking and optimizing supply chain operations.
- **Telematics:** Supports fleet management and vehicle tracking applications.

10.5 Edge Computing Solutions

1. EdgeX Foundry

Overview: EdgeX Foundry is an open-source platform that provides a flexible framework for building edge computing solutions. It supports interoperability between devices and applications at the edge.

Key Features:

- **Modularity:** Modular architecture allows for flexible deployment and customization.
- **Interoperability:** Supports multiple protocols and device interfaces.
- **Scalability:** Scalable framework for deploying edge computing applications.

Use Cases:

- **Smart Grids:** Used for managing and optimizing energy distribution in smart grids.
- **Industrial Automation:** Supports edge computing applications in manufacturing and industrial automation.

2. AWS IoT Greengrass

Overview: AWS IoT Greengrass extends AWS IoT functionality to edge devices, allowing them to act locally on the data they generate while still using the cloud for management, analytics, and durable storage.

Key Features:

- **Local Processing:** Enables local data processing and analytics on edge devices.
- **Seamless Integration:** Integration with AWS IoT Core and other AWS services.
- **Security:** Provides secure data communication and device authentication.

Use Cases:

- **Smart Agriculture:** Used for local data processing in precision farming applications.
- **Retail:** Supports in-store analytics and customer behaviour analysis.

Chapter 11: IoT Standards and Protocols

Standards and protocols are the backbone of the Internet of Things (IoT) ecosystem, ensuring seamless communication, interoperability, and security across diverse devices and platforms. This chapter delves into the key IoT standards and protocols, their importance, and their roles in fostering a unified and efficient IoT environment.

11.1 The Importance of IoT Standards and Protocols

Definition: IoT standards are established guidelines that ensure devices and systems operate in a consistent and compatible manner. Protocols are rules and conventions for data exchange between devices.

Importance:

- **Interoperability:** Standards and protocols ensure that devices from different manufacturers can communicate and work together seamlessly.
- **Scalability:** They provide a scalable framework for integrating new devices and technologies into the IoT ecosystem.
- **Security:** Standardized security protocols help protect data and devices from unauthorized access and cyber threats.
- **Reliability:** Consistent protocols enhance the reliability and stability of IoT systems by ensuring predictable and robust communication.

11.2 Key IoT Communication Protocols

1. **MQTT (Message Queuing Telemetry Transport)**

Overview: MQTT is a lightweight, publish-subscribe network protocol that transports messages between devices. It is designed for low-bandwidth, high-latency, or unreliable networks.

Key Features:

- **Lightweight:** Requires minimal network bandwidth and device resources.
- **Publish-Subscribe Model:** Decouples message producers from consumers, enhancing scalability and flexibility.
- **QoS Levels:** Provides different levels of Quality of Service (QoS) to ensure reliable message delivery.

Use Cases:

- **Smart Home:** Used for communication between home automation devices.
- **Industrial IoT:** Suitable for monitoring and controlling industrial equipment.

2. CoAP (Constrained Application Protocol)

Overview: CoAP is a specialized web transfer protocol designed for constrained devices and networks. It is built on the principles of REST (Representational State Transfer) and operates over UDP (User Datagram Protocol).

Key Features:

- **Low Overhead:** Minimal overhead makes it ideal for constrained environments.
- **RESTful Design:** Uses simple HTTP methods (GET, POST, PUT, DELETE) for interaction.
- **Efficient:** Supports multicast and asynchronous messaging.

Use Cases:

- **Smart Energy:** Used in smart grid and energy management systems.
- **Healthcare:** Suitable for remote patient monitoring and medical device communication.

3. HTTP/HTTPS

Overview: HTTP (Hypertext Transfer Protocol) and its secure counterpart HTTPS (HTTP Secure) are widely used web protocols for transferring data over the internet.

Key Features:

- **Ubiquity:** Supported by almost all web-enabled devices and applications.
- **Security:** HTTPS provides encrypted communication, ensuring data privacy and integrity.
- **Simplicity:** Well-understood and easy to implement.

Use Cases:

- **Web-Based IoT Applications:** Used for IoT devices that need to interact with web services and applications.
- **Consumer IoT:** Suitable for devices like smart appliances and wearables that connect to cloud services.

11.3 Key IoT Networking Protocols

1. Wi-Fi

Overview: Wi-Fi is a widely used wireless networking technology that allows devices to connect to the internet and each other.

Key Features:

- **High Bandwidth:** Provides high data transfer rates.
- **Wide Range:** Suitable for home and office environments.
- **Ease of Use:** Simple setup and widely supported.

Use Cases:

- **Smart Home:** Used for connecting smart home devices like cameras, thermostats, and lights.
- **Consumer Electronics:** Common in devices like smart phones, tablets, and laptops.

2. Bluetooth and Bluetooth Low Energy (BLE)

Overview: Bluetooth is a short-range wireless technology standard, while Bluetooth Low Energy (BLE) is designed for low power consumption applications.

Key Features:

- **Low Power:** BLE is energy-efficient, suitable for battery-powered devices.
- **Short Range:** Ideal for close-proximity communication.
- **Widely Supported:** Supported by most modern smart phones and devices.

Use Cases:

- **Wearables:** Used in fitness trackers, smart watches, and health monitoring devices.

- **Proximity Sensing:** Suitable for applications like beacons and asset tracking.

3. Zigbee

Overview: Zigbee is a low-power, low-data-rate wireless communication standard designed for IoT applications.

Key Features:

- **Mesh Networking:** Supports mesh networking, extending coverage and reliability.
- **Low Power:** Optimized for battery-operated devices.
- **Scalability:** Can support large networks with many devices.

Use Cases:

- **Home Automation:** Used in smart lighting, security systems, and energy management.
- **Industrial IoT:** Suitable for industrial control and monitoring applications.

11.4 IoT Data Protocols

1. OPC UA (Open Platform Communications Unified Architecture)

Overview: OPC UA is a machine-to-machine communication protocol for industrial automation developed by the OPC Foundation.

Key Features:

- **Platform Independence:** Works across different platforms and operating systems.
- **Security:** Provides robust security features, including encryption and authentication.

- **Scalability:** Suitable for small devices to enterprise systems.

Use Cases:

- **Industrial Automation:** Used for data exchange between industrial equipment and systems.
- **Energy Management:** Suitable for monitoring and controlling energy systems.

2. AMQP (Advanced Message Queuing Protocol)

Overview: AMQP is an open standard application layer protocol for message-oriented middleware.

Key Features:

- **Reliability:** Ensures reliable message delivery through various message delivery guarantees.
- **Interoperability:** Supports multiple programming languages and platforms.
- **Flexibility:** Allows for a wide range of messaging patterns.

Use Cases:

- **Financial Services:** Used in trading platforms and payment processing systems.
- **Healthcare:** Suitable for secure messaging in healthcare applications.

11.5 IoT Security Protocols

1. DTLS (Datagram Transport Layer Security)

Overview: DTLS is a protocol for securing datagram-based communications, providing the same security guarantees as TLS (Transport Layer Security).

Key Features:

- **Encryption:** Provides end-to-end encryption for secure data transmission.
- **Integrity:** Ensures data integrity through message authentication.
- **Compatibility:** Works over UDP, making it suitable for constrained networks.

Use Cases:

- **Smart Metering:** Used for secure communication in smart grid applications.
- **IoT Devices:** Suitable for securing communication between IoT devices and servers.

2. LoRaWAN (Long Range Wide Area Network)

Overview: LoRaWAN is a protocol for low-power, wide-area networks (LPWANs), designed for long-range communication with low data rates.

Key Features:

- **Long Range:** Supports communication over long distances.
- **Low Power:** Optimized for battery-powered devices.
- **Security:** Provides encryption for secure data transmission.

Use Cases:

- **Environmental Monitoring:** Used for monitoring air quality, water levels, and other environmental parameters.

- **Smart Agriculture:** Suitable for precision farming and remote monitoring of agricultural fields.

Chapter 12: The Future of IoT

The Internet of Things (IoT) is a rapidly evolving field that holds immense potential for transforming industries, improving daily life, and creating new opportunities for innovation. As technology advances and new trends emerge, the future of IoT is poised to bring about significant changes and developments. This chapter explores the key trends, advancements, and future directions in IoT, providing insights into what lies ahead for this dynamic and transformative technology.

12.1 Emerging Trends in IoT

1. Integration with Artificial Intelligence (AI) and Machine Learning (ML)

Overview: AI and ML technologies are becoming increasingly integral to IoT systems, enhancing their capabilities and applications. By leveraging AI and ML, IoT devices can process and analyze data more intelligently, making real-time decisions and predictions.

Key Trends:

- **Predictive Analytics:** AI algorithms can analyze data from IoT devices to predict future trends, maintenance needs, and anomalies.
- **Automated Decision-Making:** IoT systems can automate responses and actions based on AI-driven insights, improving efficiency and reducing human intervention.
- **Enhanced Personalization:** AI can tailor IoT applications to individual user preferences and behaviours, creating more personalized experiences.

Implications:

- **Smart Homes:** AI-powered smart home devices can learn user preferences and adjust settings automatically.
- **Industrial Automation:** AI can optimize manufacturing processes, predict equipment failures, and improve quality control.

2. 5G and Advanced Connectivity

Overview: 5G technology is set to revolutionize IoT connectivity by providing faster data transfer speeds, lower latency, and greater network capacity. This will enable a new generation of IoT applications that require high-speed, reliable communication.

Key Trends:

- **Enhanced Bandwidth:** 5G offers significantly higher bandwidth compared to previous generations, supporting more connected devices and higher data throughput.
- **Low Latency:** Reduced latency enables real-time communication and faster response times for IoT applications.
- **Network Slicing:** 5G allows for network slicing, which can create virtual networks optimized for specific IoT use cases.

Implications:

- **Smart Cities:** 5G will support advanced smart city applications like real-time traffic management, autonomous vehicles, and smart infrastructure.
- **Healthcare:** Faster connectivity will enable more reliable remote patient monitoring and telemedicine services.

3. Edge Computing and Local Processing

Overview: Edge computing is gaining traction as a means of processing data closer to the source, reducing latency, and improving the performance of IoT applications. By moving data processing to the edge of the network, IoT systems can respond more quickly to local events.

Key Trends:

- **Real-Time Analytics:** Edge computing enables real-time data processing and analytics, reducing the need to send data to centralized cloud servers.
- **Improved Reliability:** Local processing can enhance system reliability by reducing dependency on network connectivity.
- **Reduced Bandwidth Usage:** Processing data locally reduces the amount of data transmitted over the network, lowering bandwidth costs.

Implications:

- **Industrial IoT:** Edge computing can enhance real-time monitoring and control in industrial environments, improving operational efficiency.
- **Smart Devices:** Edge processing can enable more responsive and intelligent smart devices, such as security cameras and wearables.

12.2 Innovations and Technological Advancements

1. Blockchain for IoT Security

Overview: Blockchain technology is being explored as a solution to enhance the security and integrity of IoT systems. By providing a decentralized and tamper-proof ledger, blockchain can address various security challenges in IoT.

Key Trends:

- **Decentralized Security:** Blockchain can eliminate single points of failure by distributing security across a network of nodes.
- **Data Integrity:** Blockchain ensures data integrity and traceability, making it difficult for unauthorized parties to alter or tamper with data.
- **Smart Contracts:** Blockchain enables the use of smart contracts, which can automate and enforce agreements between IoT devices and systems.

Implications:

- **Supply Chain Management:** Blockchain can enhance transparency and traceability in supply chains, improving security and efficiency.
- **IoT Security:** Blockchain can strengthen security protocols for IoT devices, reducing the risk of cyber attacks and data breaches.

2. Advancements in Sensor Technology

Overview: Sensor technology is continuously evolving, with advancements leading to more accurate, compact, and energy-efficient sensors. These improvements are driving the next generation of IoT applications.

Key Trends:

- **Miniaturization:** Smaller sensors allow for more compact and versatile IoT devices.
- **Increased Sensitivity:** Improved sensor sensitivity enables more precise measurements and data collection.
- **Energy Efficiency:** Advances in sensor technology are reducing power consumption, extending the battery life of IoT devices.

Implications:

- **Wearables:** Enhanced sensors in wearables can provide more accurate health and fitness tracking.
- **Environmental Monitoring:** Improved sensors can offer better data for monitoring environmental conditions and managing resources.

12.3 Challenges and Considerations

1. Data Privacy and Ethics

Overview: As IoT devices collect and analyze vast amounts of personal data, data privacy and ethical considerations are becoming increasingly important. Ensuring that data is handled responsibly and transparently is critical for maintaining user trust.

Key Trends:

- **Data Protection Regulations:** Compliance with data protection regulations such as GDPR and CCPA is essential for safeguarding user privacy.
- **Ethical Data Use:** Organizations must consider ethical implications related to data collection, usage, and sharing.
- **User Consent:** Obtaining explicit user consent for data collection and usage is crucial for maintaining transparency.

Implications:

- **Consumer Trust:** Protecting user privacy and ensuring ethical data practices are essential for building and maintaining consumer trust in IoT solutions.

- **Regulatory Compliance:** Organizations must stay informed about evolving data protection regulations and ensure compliance.

2. Integration Challenges

Overview: As IoT systems become more complex, integrating various devices, platforms, and technologies presents significant challenges. Ensuring seamless integration and interoperability is crucial for the success of IoT deployments.

Key Trends:

- **Complex Ecosystems:** IoT ecosystems often involve multiple devices, platforms, and protocols, making integration complex.
- **Standardization:** The lack of standardized protocols and frameworks can hinder interoperability and integration efforts.
- **Vendor Lock-In:** Proprietary solutions may create dependencies on specific vendors, limiting flexibility and scalability.

Implications:

- **System Design:** Careful design and planning are required to address integration challenges and ensure seamless operation across IoT systems.
- **Interoperability:** Promoting the adoption of open standards and interoperability frameworks can facilitate smoother integration and collaboration.

12.4 The Road Ahead

Overview: The future of IoT is characterized by continuous innovation and rapid technological advancements. To stay competitive and capitalize on emerging opportunities, organizations must remain agile and adaptable.

Key Directions:

- **Continuous Innovation:** Embracing new technologies and trends will be crucial for driving innovation and creating value in IoT applications.
- **Collaboration:** Collaboration between industry stakeholders, technology providers, and regulators will be essential for addressing challenges and advancing the IoT ecosystem.
- **User-Centric Design:** Focusing on user needs and preferences will drive the development of more intuitive and effective IoT solutions.

Implications:

- **Strategic Planning:** Organizations should develop strategic plans to navigate the evolving IoT landscape and leverage emerging technologies.
- **Investment in Research:** Investing in research and development will be key to staying ahead of technological trends and driving future growth in IoT.

Chapter 13: IoT Case Studies

To understand the practical applications and impact of IoT technologies, it is essential to explore real-world case studies. These examples demonstrate how IoT is transforming various industries by enhancing operational efficiency, improving customer experiences, and driving innovation. This chapter presents a selection of case studies that highlight the diverse applications of IoT across different sectors, including healthcare, manufacturing, transportation, agriculture, and smart cities.

13.1 Healthcare

Case Study: Remote Patient Monitoring by Medtronic

Overview: Medtronic, a leading medical technology company, has developed a remote patient monitoring solution to improve the management of chronic conditions such as diabetes and heart disease. The system leverages IoT devices to continuously monitor patient health metrics and transmit data to healthcare providers.

Implementation:

- **Devices:** Patients use wearable sensors and glucose monitors that track vital signs, blood glucose levels, and heart activity.
- **Data Transmission:** The collected data is transmitted in real-time to a cloud-based platform via cellular or Wi-Fi networks.
- **Analytics:** Advanced analytics and machine learning algorithms analyze the data to detect anomalies and predict potential health issues.

Benefits:

- **Improved Patient Outcomes:** Real-time monitoring allows for early intervention and personalized treatment plans.

- **Reduced Hospital Visits:** Remote monitoring reduces the need for frequent in-person visits, saving time and resources.
- **Enhanced Patient Engagement:** Patients can track their health data and receive feedback, leading to better self-management.

Challenges:

- **Data Security:** Ensuring the security and privacy of sensitive health data is critical.
- **Device Integration:** Integrating different types of medical devices and ensuring compatibility can be complex.

13.2 Manufacturing

Case Study: Predictive Maintenance by Siemens

Overview: Siemens, a global industrial manufacturing company, has implemented a predictive maintenance solution in its factories to optimize equipment performance and reduce downtime. The system uses IoT sensors and data analytics to monitor machinery and predict maintenance needs.

Implementation:

- **Sensors:** IoT sensors are installed on critical equipment to monitor parameters such as temperature, vibration, and pressure.
- **Data Collection:** The sensors continuously collect data and send it to a centralized analytics platform.
- **Predictive Analytics:** Machine learning algorithms analyze the data to predict equipment failures and recommend maintenance actions.

Benefits:

- **Reduced Downtime:** Predictive maintenance minimizes unplanned downtime by addressing issues before they cause failures.

- **Cost Savings:** Early detection of potential problems reduces repair costs and extends equipment lifespan.
- **Enhanced Efficiency:** Improved equipment reliability leads to higher production efficiency and quality.

Challenges:

- **Data Management:** Handling and analyzing large volumes of data from numerous sensors can be challenging.
- **Integration:** Integrating IoT solutions with existing manufacturing systems and processes requires careful planning.

13.3 Transportation

Case Study: Fleet Management by Geotab

Overview: Geotab, a leading fleet management solutions provider, uses IoT technology to optimize the management of vehicle fleets. The solution offers real-time tracking, monitoring, and analytics for fleet operators.

Implementation:

- **Telematics Devices:** Vehicles are equipped with Telematics devices that track location, speed, fuel consumption, and engine performance.
- **Data Transmission:** Data is transmitted to a cloud-based platform for analysis and visualization.
- **Analytics:** The platform provides insights into fleet performance, driver behaviour, and vehicle maintenance needs.

Benefits:

- **Operational Efficiency:** Fleet operators can optimize routes, reduce fuel consumption, and improve overall efficiency.

- **Safety:** Real-time monitoring of driver behaviour and vehicle health enhances safety and reduces the risk of accidents.
- **Cost Savings:** Improved maintenance scheduling and route optimization result in cost savings.

Challenges:

- **Data Privacy:** Ensuring the privacy and security of vehicle and driver data is essential.
- **Connectivity:** Reliable connectivity is required for real-time data transmission and monitoring.

13.4 Agriculture

Case Study: Precision Farming by John Deere

Overview: John Deere, a leading agricultural equipment manufacturer, has developed precision farming solutions that leverage IoT technology to enhance agricultural practices. The solution uses IoT sensors and data analytics to optimize crop management and improve yields.

Implementation:

- **Sensors:** IoT sensors are deployed in fields to monitor soil conditions, weather, and crop health.
- **Data Collection:** Sensors collect data on soil moisture, nutrient levels, and weather conditions, transmitting it to a central platform.
- **Analytics:** Data is analyzed to provide actionable insights for irrigation, fertilization, and pest control.

Benefits:

- **Increased Yields:** Precision farming techniques optimize resource usage, leading to higher crop yields.

- **Resource Efficiency:** Efficient use of water, fertilizers, and pesticides reduces waste and environmental impact.
- **Cost Savings:** Improved management practices lower operational costs and increase profitability.

Challenges:

- **Data Accuracy:** Ensuring the accuracy of sensor data and analytics is critical for effective decision-making.
- **Infrastructure:** Reliable infrastructure is needed for data transmission and processing in rural areas.

13.5 Smart Cities

Case Study: Smart Traffic Management by Barcelona

Overview: Barcelona, Spain, has implemented a smart traffic management system to improve urban mobility and reduce traffic congestion. The system uses IoT sensors and data analytics to monitor and manage traffic flow in real-time.

Implementation:

- **Traffic Sensors:** IoT sensors are installed at key intersections and along major roads to monitor traffic conditions.
- **Data Integration:** Data from sensors is integrated with a central traffic management platform.
- **Real-Time Control:** The platform analyzes traffic patterns and adjusts traffic signals to optimize flow and reduce congestion.

Benefits:

- **Reduced Congestion:** Real-time traffic management reduces congestion and improves travel times.

- **Environmental Impact:** Efficient traffic flow reduces vehicle emissions and contributes to a cleaner environment.
- **Enhanced Safety:** Improved traffic management reduces the risk of accidents and enhances road safety.

Challenges:

- **Data Privacy:** Managing and protecting data collected from traffic sensors is essential.
- **System Integration:** Integrating IoT solutions with existing traffic management infrastructure requires coordination.

Chapter 14: Ethical and Societal Implications of IoT

The rapid proliferation of Internet of Things (IoT) technologies has brought about significant advancements and conveniences in various aspects of daily life and industry. However, as with any transformative technology, IoT also presents a range of ethical and societal challenges. This chapter explores the ethical considerations and societal impacts of IoT, examining issues related to privacy, security, inequality, and the broader implications for human behaviour and social structures.

14.1 Privacy Concerns

1. Data Collection and Surveillance

Overview: IoT devices continuously collect data from users and environments, leading to concerns about the extent of surveillance and data collection.

Key Issues:

- **Intrusiveness:** Continuous data collection from personal devices, such as smart home assistants and wearable, can be perceived as intrusive.
- **Data Ownership:** Questions arise about who owns the data collected by IoT devices and how it is used by manufacturers and third parties.
- **Surveillance:** The potential for pervasive surveillance by governments or corporations raises concerns about privacy and civil liberties.

Implications:

- **User Consent:** Ensuring informed user consent for data collection and usage is essential for maintaining trust and privacy.
- **Transparency:** Companies must be transparent about data practices and provide clear information on data usage policies.

2. **Data Security and Breaches**

Overview: IoT devices often store and transmit sensitive information, making them potential targets for cyber attacks and data breaches.

Key Issues:

- **Vulnerability:** Many IoT devices have security vulnerabilities that can be exploited by hackers.
- **Data Breaches:** Compromised data can lead to identity theft, financial loss, and other privacy violations.
- **Security Standards:** The lack of standardized security protocols for IoT devices increases the risk of breaches.

Implications:

- **Robust Security Measures:** Implementing strong security measures and regular updates is crucial for protecting IoT devices and user data.
- **Industry Standards:** Developing and adhering to security standards can help mitigate risks and enhance overall device security.

14.2 Ethical Considerations

1. **Bias and Fairness**

Overview: AI and machine learning algorithms used in IoT systems can perpetuate or exacerbate existing biases, leading to ethical concerns about fairness and discrimination.

Key Issues:

- **Algorithmic Bias:** IoT systems that rely on AI may unintentionally reinforce biases present in the training data, leading to unfair outcomes.
- **Discrimination:** Biased algorithms can result in discriminatory practices, particularly in areas like hiring, lending, and law enforcement.

Implications:

- **Bias Mitigation:** Developing and implementing strategies to detect and mitigate bias in algorithms is essential for ensuring fairness.
- **Inclusive Design:** Designing IoT systems with diversity and inclusivity in mind can help reduce the risk of biased outcomes.

2. Autonomy and Control

Overview: IoT technologies can impact personal autonomy by automating decisions and actions, potentially reducing human control over certain aspects of life.

Key Issues:

- **Decision-Making:** Automated systems may make decisions without human input, raising concerns about control and accountability.
- **Dependence:** Increased reliance on IoT technologies can lead to diminished human skills and decision-making abilities.

Implications:

- **Human Oversight:** Ensuring that automated systems include options for human oversight and intervention is important for maintaining control.
- **User Empowerment:** Providing users with clear choices and control over IoT functionalities can help preserve autonomy.

14.3 Societal Impacts

1. Digital Divide

Overview: The adoption of IoT technologies can exacerbate existing inequalities, creating a digital divide between those with access to technology and those without.

Key Issues:

- **Access:** Disparities in access to IoT technologies can widen the gap between different socioeconomic groups.
- **Affordability:** The cost of IoT devices and services may be prohibitive for some individuals or communities.

Implications:

- **Bridging the Divide:** Initiatives to increase access to IoT technologies and improve affordability can help address inequalities.
- **Inclusive Design:** Designing IoT solutions that consider diverse needs and contexts can contribute to greater inclusivity.

2. Behavioural Changes

Overview: The pervasive presence of IoT technologies can influence human behaviour and social interactions, leading to changes in how people live and communicate.

Key Issues:

- **Behavioural Impact:** IoT devices can affect daily routines, communication patterns, and social interactions.
- **Privacy Concerns:** The constant monitoring and data collection may lead to altered behaviour and heightened self-consciousness.

Implications:

- **Awareness:** Raising awareness about the potential behavioural impacts of IoT technologies can help individuals make informed choices.
- **Balance:** Striking a balance between the benefits of IoT and the need for privacy and autonomy is crucial for maintaining healthy behaviours.

14.4 Regulatory and Policy Considerations

1. Regulation of IoT Devices

Overview: The rapid growth of IoT technologies necessitates the development of regulations and policies to address ethical and security concerns.

Key Issues:

- **Regulatory Frameworks:** Existing regulations may not adequately address the unique challenges posed by IoT technologies.

- **Global Standards:** The need for global standards and regulations to ensure consistency and protection across borders.

Implications:

- **Policy Development:** Governments and regulatory bodies need to develop comprehensive policies that address the ethical and security aspects of IoT.
- **International Cooperation:** Collaboration between countries and international organizations can help create cohesive and effective regulatory frameworks.

2. Ethical Guidelines for IoT Development

Overview: Establishing ethical guidelines for the development and deployment of IoT technologies can help address concerns related to privacy, security, and fairness.

Key Issues:

- **Ethical Standards:** Developing and adopting ethical standards for IoT development can guide industry practices and ensure responsible innovation.
- **Stakeholder Involvement:** Engaging various stakeholders, including users, developers, and policymakers, in the creation of ethical guidelines is important for comprehensive oversight.

Implications:

- **Best Practices:** Promoting best practices in IoT development can help mitigate ethical concerns and enhance overall trust in the technology.

- **Education and Training:** Providing education and training on ethical considerations can support responsible development and usage of IoT technologies.

Chapter 15: Summary and Future Directions

As we conclude our exploration of the Internet of Things (IoT), it is essential to review the key insights and themes discussed throughout the book and consider the future directions of this transformative technology. This final chapter provides a summary of the key takeaways, reflects on the implications of IoT advancements, and outlines the potential future trends and challenges that may shape the evolution of IoT.

15.1 Summary of Key Takeaways

1. The Evolution of IoT

IoT technology has evolved from its early beginnings to become a critical component of modern life and industry. From simple sensor networks to sophisticated, interconnected systems, IoT has transformed the way we interact with technology and the world around us. Key advancements include the integration of AI and machine learning, the deployment of 5G networks, and the development of edge computing solutions.

2. Applications Across Industries

IoT has demonstrated its value across a wide range of industries, including healthcare, manufacturing, transportation, agriculture, and smart cities. Case studies highlighted the diverse applications of IoT, such as remote patient monitoring, predictive maintenance, fleet management, precision farming, and smart traffic management. These applications showcase IoT's ability to enhance operational efficiency, improve quality of life, and drive innovation.

3. Ethical and Societal Implications

The widespread adoption of IoT technology brings with it a range of ethical and societal concerns. Issues related to privacy, data security, algorithmic bias, and the digital divide must be carefully addressed to ensure that IoT benefits are distributed equitably and responsibly. Developing robust regulatory frameworks, ethical guidelines, and inclusive practices is crucial for navigating these challenges.

15.2 Implications for the Future

1. Continued Innovation and Integration

The future of IoT will be characterized by ongoing innovation and integration with other emerging technologies. Advances in AI, 5G, edge computing, and Blockchain are expected to drive the next generation of IoT applications and solutions. This integration will enable more intelligent, responsive, and efficient systems that can address complex challenges and create new opportunities.

2. Expanding IoT Ecosystems

As IoT ecosystems continue to expand, interoperability and standardization will become increasingly important. The development of universal standards and protocols will facilitate seamless integration between devices, platforms, and services, enabling more cohesive and effective IoT solutions. Collaboration between industry stakeholders, technology providers, and regulators will be essential for achieving this goal.

3. Addressing Ethical and Societal Challenges

Addressing ethical and societal challenges will remain a critical focus for the future of IoT. Ensuring data privacy, security, and fairness will require ongoing efforts to develop and implement best practices, regulations, and guidelines. By proactively addressing these concerns, we can build trust in IoT technologies and ensure that their benefits are realized in a responsible and equitable manner.

15.3 Future Directions

1. Smart and Connected Environments

The future of IoT will likely see the development of increasingly smart and connected environments, including homes, cities, and workplaces. These environments will leverage IoT technologies to create more efficient, responsive, and sustainable systems. Innovations such as smart grids, autonomous vehicles, and intelligent infrastructure will shape the future landscape of IoT.

2. Personalized and Adaptive Solutions

As IoT technologies advance, there will be a greater emphasis on creating personalized and adaptive solutions that cater to individual needs and preferences. AI-driven insights and machine learning algorithms will enable IoT systems to deliver more tailored experiences, from personalized healthcare interventions to customized smart home environments.

3. Ethical AI and Responsible Development

The integration of AI with IoT will necessitate a focus on ethical AI and responsible development practices. Ensuring that AI algorithms are transparent, unbiased, and aligned with ethical principles will be essential for maintaining trust and ensuring that IoT solutions are used for positive and equitable purposes.

4. Global Collaboration and Regulation

Global collaboration and regulation will play a crucial role in shaping the future of IoT. As IoT technologies continue to evolve and expand across borders, international cooperation will be necessary to address common challenges, establish global standards, and promote responsible development and usage.

Chapter 16: Emerging Trends and Technologies in IoT

As the Internet of Things (IoT) continues to evolve, several emerging trends and technologies are set to shape its future. This chapter delves into the latest advancements and innovations in IoT, exploring how these trends are likely to influence the development and deployment of IoT solutions. We will examine the impact of advancements in areas such as artificial intelligence (AI), edge computing, 5G, blockchain, and the Internet of Everything (IoE). Understanding these trends will provide insights into the future direction of IoT and its potential to revolutionize various sectors.

16.1 Artificial Intelligence (AI) and Machine Learning

1. Enhanced Analytics and Insights

Overview: Artificial intelligence and machine learning are transforming IoT by providing advanced analytics and insights. AI algorithms can process vast amounts of data generated by IoT devices, enabling more sophisticated analysis and decision-making.

Key Developments:

- **Predictive Analytics:** AI-driven predictive analytics can forecast trends, detect anomalies, and anticipate issues before they arise.
- **Natural Language Processing (NLP):** NLP allows for more intuitive interactions with IoT systems, such as voice-controlled smart home devices.

Impact:

- **Improved Efficiency:** AI enhances the efficiency and effectiveness of IoT solutions by providing actionable insights and automating complex processes.
- **Personalization:** AI enables more personalized experiences by analyzing user behaviour and preferences to tailor services and recommendations.

Challenges:

- **Data Quality:** The effectiveness of AI relies on the quality and accuracy of the data collected by IoT devices.
- **Ethical Considerations:** Ensuring transparency and fairness in AI algorithms is crucial for maintaining trust and avoiding bias.

16.2 Edge Computing

1. Processing Data Locally

Overview: Edge computing involves processing data closer to the source, reducing latency and bandwidth usage. This approach is particularly valuable for IoT applications that require real-time processing and low latency.

Key Developments:

- **Edge Devices:** Devices with local processing capabilities, such as smart sensors and gateways, can perform computations and analytics on-site.
- **Reduced Latency:** By processing data locally, edge computing minimizes the delay between data collection and action, enhancing the responsiveness of IoT systems.

Impact:

- **Real-Time Processing:** Edge computing supports real-time data processing and decision-making, which is critical for applications like autonomous vehicles and industrial automation.
- **Bandwidth Savings:** Local processing reduces the amount of data transmitted to centralized servers, saving bandwidth and improving overall network performance.

Challenges:

- **Security:** Ensuring the security of data and computations performed on edge devices is essential to protect against potential vulnerabilities.
- **Management:** Managing and maintaining a large number of edge devices can be complex and require new strategies and tools.

16.3 5G Technology

1. High-Speed Connectivity

Overview: 5G technology promises to revolutionize IoT with its high-speed, low-latency connectivity. This next-generation network offers significant improvements over previous generations, enabling more advanced IoT applications.

Key Developments:

- **Increased Bandwidth:** 5G provides higher data transfer speeds, allowing for faster communication between IoT devices and systems.
- **Low Latency:** The reduced latency of 5G enhances the responsiveness of IoT applications, such as remote control and real-time monitoring.

Impact:

- **Enhanced Performance:** 5G enables more reliable and high-performance IoT solutions, supporting applications that require real-time data and high bandwidth.
- **Scalability:** The capacity of 5G networks supports the growing number of IoT devices and use cases, facilitating the expansion of IoT ecosystems.

Challenges:

- **Infrastructure:** Implementing 5G infrastructure requires significant investment and development, particularly in areas with limited connectivity.
- **Interference:** Managing potential interference and ensuring network reliability in densely populated areas is a key consideration.

16.4 Blockchain Technology

1. Secure and Transparent Transactions

Overview: Blockchain technology offers a decentralized and secure way to manage data and transactions. Its integration with IoT can enhance data integrity, security, and transparency.

Key Developments:

- **Smart Contracts:** Blockchain-based smart contracts automate and enforce agreements between IoT devices and systems without the need for intermediaries.
- **Data Integrity:** Blockchain's immutable ledger ensures that data collected by IoT devices is secure and tamper-proof.

Impact:

- **Enhanced Security:** Blockchain provides a secure framework for managing IoT data and transactions, reducing the risk of tampering and fraud.
- **Transparency:** The transparency of blockchain transactions enables greater visibility and trust in IoT systems.

Challenges:

- **Scalability:** The scalability of blockchain solutions must be addressed to handle the large volumes of data generated by IoT devices.
- **Integration:** Integrating blockchain with existing IoT systems and technologies requires careful planning and development.

16.5 Internet of Everything (IoE)

1. Interconnected Ecosystems

Overview: The Internet of Everything (IoE) extends the concept of IoT to include not only devices but also people, processes, and data. IoE envisions a fully interconnected ecosystem where all elements work together seamlessly.

Key Developments:

- **Holistic Integration:** IoE aims to integrate devices, people, and processes into a cohesive system that enhances overall efficiency and functionality.
- **Data Utilization:** By connecting all aspects of the ecosystem, IoE enables more comprehensive and effective use of data to drive insights and actions.

Impact:

- **Enhanced Collaboration:** IoE fosters collaboration between devices, systems, and individuals, leading to more synchronized and efficient operations.
- **Innovation:** The interconnected nature of IoE drives innovation by enabling new applications and use cases that leverage the full potential of connected elements.

Challenges:

- **Complexity:** Managing and coordinating a complex ecosystem of interconnected elements can be challenging and requires robust systems and protocols.
- **Privacy:** Ensuring the privacy and security of data in a highly interconnected environment is crucial for maintaining trust and protecting sensitive information.

Conclusion

The landscape of IoT is rapidly evolving, driven by advancements in AI, edge computing, 5G, blockchain, and the broader concept of the Internet of Everything. These emerging trends and technologies are set to transform the way IoT solutions are developed, deployed, and utilized. As we look to the future, it is essential to stay informed about these developments and consider their implications for various industries and applications.

By understanding and embracing these trends, stakeholders can harness the full potential of IoT while addressing the challenges and opportunities that arise. The continued evolution of IoT will undoubtedly bring about new innovations and possibilities, shaping the future of technology and its impact on our world.

www.ingramcontent.com/pod-product-compliance
Lightning Source LLC
Chambersburg PA
CBHW071950210526
45479CB00003B/876